Primary Word Problems

Book 2, 2nd Edition

Problem-Solving Ideas for Grades 3-5

Reaching Your Dream Learning Cent

DR. NOW'S MATH

MAKING THE IMPOSSIBLE, POSSIBLE!

by **Dr. Now**

MARTINA PUBLISHING, INC

MsLiteracy.com

Martina Publishing, Inc., PO Box 1216, Walterboro, SC 29488
www.MsLiteracy.com

PRIMARY WORD PROBLEMS: Problem-Solving Ideas for Grades 3-5
BOOK 2, Second Edition
by Rosemary Simmons-Brown (a.k.a. Dr. Now)

SUMMARY:
This engaging and educational math book is crafted to enhance the problem-solving abilities of children aged 8 to 12. The second edition presents various word problems that promote critical thinking and help establish a solid mathematical foundation, featuring a fresh design. Perfect for home or classroom use, this resource aims to equip young learners for future academic challenges through enjoyable and interactive math games and exercises.

Target Audience: Ages 8 - 12

ISBN-13: 978-1-957645-23-0
Library of Congress Control Number: 2025915425

1. Basic Mathematics--Word Problems. 2. Math Games--Juvenile Literature. 3. Critical Thinking. 4. Problem-Solving--Math Concepts. I. Title]

Dedication

The book is dedicated to my dad, Willie Kizer Crum, who truly believed in me.

Day 1:

Your favorite uncle is driving in from Del City, Oklahoma, to your home next week. He's coming from Del City, Oklahoma, to Walterboro, South Carolina. Your uncle lives 1,083.6 miles away from you. Approximately, how many hours will it take your uncle to get home?

Workspace

Day 2:

You plan to share a pizza with two friends. Just before you cut the pizza, a fourth friend comes to join you. Show and explain what you will do so that everyone has an equal slice of the pizza.

Workspace

Day 3:

Your favorite uncle gave you $75.00. You are going to spend every cent. Your choices are:

a pad case, $31.00	a bowling ticket, $24.00
an aquarium, $29.00	a video game $44.00

What will you have to purchase to come as close as possible to using all your $75.00?

Workspace

Day 4:

Fifteen children are in Dr. Now's class. Nine of the children are girls. How many boys are in Dr. Now's class?

There are ___ boys in Dr. Now's class.

Workspace

Day 5:

If Saturday school begins at 9:00 am and ends three hours later, what time will Saturday school end?

Saturday School will end at ____________.

Workspace

Day 6:

There were 14 children at a party. There were 36 cupcakes. If each child ate two cupcakes, how many cupcakes were left?

There were _____ cupcakes left.

Workspace

Day 7:

I can pick 50 tomatoes in 2 hours.
How many tomatoes can I pick in 8 hours?

I can pick _______ tomatoes in 8 hours.

Workspace

Day 8:

There are 289 sheep in the field. Each sheep has four legs.
How many legs do the 289 sheep have altogether?

There are 289 sheep in the field with ____________ legs altogether.

Workspace

Day 9:

There are 165 calories in 15 potato chips.
How many calories are there in each potato chip?

Each potato chip has ______ calories.

Workspace

Day 10:

A daycare center ordered 244 new blankets.
The blankets will be delivered in 4 equal shipments.
How many blankets will be in each shipment?

There will be _____ blankets in each shipment.

Workspace

Day 11:

The marching band has 200 students.
How many rows of 10 students can be formed in each row?

There are ____ rows of students in the marching band.

Workspace

Day 12:

The soda factory filled 7,432 bottles in 2 hours.
How many bottles were filled every half hour (30 minutes)?

Workspace

Day 13:

Mr. Simmons has 588 cement blocks. He is building a storage shed. How many rows of 12 cement blocks can he lay for the storage shed?

Workspace

Day 14:

Ethan received a new video game for his birthday.
He scored 4 points on the first level, 16 points on the second level, and 64 points on the third level. If this pattern continues, how many points will Ethan score on the fourth level?

Workspace

Day 15:

Malachi has a marble collection and likes to display his collection in trays. He puts 3 marbles in the first tray, 5 marbles in the second tray, 8 marbles in the third tray, and 12 marbles in the fourth tray. If this pattern continues, how many marbles will Malachi display in the fifth tray?

Workspace

Day 16:

While putting potatoes in piles for the potato bank, Uncle George put 1 potato in the first pile, 2 potatoes in the second pile, 5 potatoes in the third pile, and 10 potatoes in the fourth pile. If he continues this pattern, how many potatoes will be put in the fifth pile?

Workspace

Day 17:

While sorting some hair barrettes, Elizabeth drops 21 barrettes in the first jar, 28 barrettes in the second jar, 35 barrettes in the third jar, 42 barrettes in the fourth jar, and 49 barrettes in the fifth jar. If she continues this pattern, how many barrettes will Elizabeth drop in the sixth jar?

Workspace

Day 18:

Approximately how many centimeters equal(s) 2 inches?

Workspace

Day 19:

Approximately how many quarts is equal to 5 liters?

Workspace

Day 20:

Myron has a piece of green yarn and a piece of yellow yarn. The piece of green yarn is 5 inches shorter than the piece of yellow yarn. Myron also has 7 inches of scotch tape. The pieces of yellow yarn are 29 inches long. How many inches of yarn does Myron have in all?

Workspace

Day 21:

How many years are in a fourth of a century?

Workspace

Day 22:

How many minutes are in 1/10 of an hour?

Workspace

Day 23:

How many years are in ½ of a decade?

Workspace

Day 24:

What fraction of a day is 4 hours?

Workspace

Day 25:

Gabrielle has 10 pink ribbons. She has 8 fewer green ribbons than pink ribbons. She has 16 more purple ribbons than green ribbons. How many ribbons does Gabrielle have in all?

Workspace

Day 26:

Ethan saved $54.00. He earned $35.00 for cutting the lawn. Then he spent $10.00 on a pair of goggles, $15.00 on an ipad cover, $14.75 for a sports cap, and $25.00 for a blue scarf. How much money does Ethan have left?

Workspace

Day 27:

In Michael's desk drawer, there are 14 yellow markers. There are 4 more black markers than yellow markers, and there are 11 more green markers than black markers. How many markers are in Michael's desk drawer altogether?

Workspace

Day 28:

Jason planted four types of fruit trees in his vineyard. He planned 37 of each type of fruit. How many fruit trees did Jason plant in his vineyard?

Workspace

Day 29:

The Simmons' Explorer has 816 campers. If each cabin can hold 8 campers, how many cabins will they need?

Workspace

Day 30:

Elizabeth was collecting sparkling marbles. She got 23 marbles in a fish bowl, and Emmanuel gave her 8 marbles. Her dad gave her 14 marbles. Elizabeth gave 17 marbles to her cousin Alex. How many marbles did Elizabeth have left?

Workspace

Day 31:

George wants to triple the size of his work area. He has an area of 123 feet. How much work area does George want?

- Do you understand the question?
- Make a plan to solve the problem.
- Carry out your plan to solve the problem.
- Check your answer.

Workspace

Day 32:

Rosie is putting a border of flowers on a rectangular driveway. The drive is 40 feet long and 26 feet wide. How long does the border of flowers need to be?

- Do you understand the question?
- Make a plan to solve the problem.
- Carry out you plan to solve the problem.
- Check your answer

Workspace

Day 33:

Janee is putting a border around her portfolio cover.
The cover is 18 inches tall and 12 inches wide.
How much border does Janee need?

Workspace

Day 34:

Your mother needs to put a square fence around her rose garden with a side length of 13 yards. How many yards of fence does she need?

Workspace

Day 35:

A daycare center teacher ordered 69 boxes of popsicle sticks. There were 885 popsicle sticks in each box. How many popsicle sticks did the daycare center order in all?

Workspace

Day 36:

The cooks at the shelter need 654 potatoes. There are 6 potatoes in each sack. How many sacks of potatoes do they need?

Workspace

Day 37:

The school principal has $6,918 to spend on replacing tiles for the classroom. If the cost of each tile is $3.00 how many tiles can the principal buy?

Workspace

Day 38:

A district of teachers needs to go to a workshop in a nearby city. There are 780 teachers in the district. If each passenger van carries 15 teachers, how many passenger vans does the district need?

Workspace

Day 39:

A cheerleading squad has \$829.00 to spend on buttons to give to their fans. The cost of each button is \$0.85. Approximately how many buttons can the squad purchase?

The squad can purchase approcimately ____________ buttons for their fans.

Workspace

Day 40:

At the Carolina Fair, a group of 554 seniors want to ride the giant Tagmaball. If each booth on the Tagmaball can hold 9 seniors in each car, approximately how many cars will the seniors need?

The seniors will need approximately _________ cars.

Workspace

Day 41:

Your dad spent $287.00 on a diamond necklace and $878.00 on a set of diamond earrings. How much money did he spend in all for your mother's birthday jewelry?

Workspace

Day 42:

Before lunch, Jason has a science class for 1 hour and 40 minutes. Lunch lasts 45 minutes and ends at 1:00 pm. What time does his science class start?

Workspace

Day 43:

You need to drink 7 ¼ cup of water each day and ¾ cups of milk each day. Approximately, how much liquid do you have to drink daily?

Workspace

Day 44:

The class has 24 students, and ¾ of them are boys. how many are boys in the class?

Workspace

Day 45:

The cook made 48 cupcakes. One-eighth of the cupcakes have green icing. How many of the cupcakes are green?

Workspace

Day 46:

Three-sixths of the 18 children on the playground are playing kickball. How many children are playing kickball?

Workspace

Day 47:

Thirty-six students are in the cafeteria eating lunch. Three-fourths of the students are eating cafeteria lunch. How many students are eating the cafeteria lunch?

Workspace

Day 48:

There are 50 people waiting for the Greyhound bus. One-fifth of them have two or more pieces of luggage. How many people have more than two pices of luggage?

Workspace

Day 49:

One-fifth of the 895 students at school wear glasses. How many students wear glasses?

Workspace

Day 50:

There are 750 bouses in our community. Two-fifths of the house are made of bricks. How many houses are made of bricks?

Workspace

Day 51:

Renting a bike at the amusement park costs $12.00 per hour. A family of four rented 4 bikes, How much will they pay to rent the bikes for 6 hours?

Workspace

Day 52:

Jason earned $25.00 for grooming the dog. He earned $35.00 for cutting the lawn. His brother Micheal earned $65.00 for replacing a ceiling fan. How much money did Jason earn in all?

Workspace

Day 53:

Janee' walks for 30 minutes each day. She does aerobics for 25 minutes each day. How many minutes does Janee' walk and do aerobics in 7 days.

Workspace

Day 54:

Tickets to the amusement park cost $12.00 for children and $15.00 for adults. How much will it cost for 6 adults and 9 children to attend the amusement park?

Talk with partner the correct order of the operations you need to use to solve the problem: multiply and then subtract; subtract and then multiply; multiply and then add; or add then multiply.

Workspace

Day 55:

It takes Michael 53 minutes to walk from his house to the store. If Michael needs to be at the store for work at 8:30 a.m., what time does he need to leave his house?

Micheal needs to leave his house by __________ a.m. to get to work on time.

Workspace

Day 56:

Mrs. Simmons bought a purse and a pair of shoes at the Department Store. She paid $59.00 for the shoes and $29.99 for the purse. The clerk gave her $1.01 change. How much money did she give the clerk?

Workspace

Day 57:

Reduce the fraction to its lowest term: 4/12

Workspace

Day 58:

Reduce the fraction to its lowest term: 6/8

Workspace

Day 59:

What is the area of the rectangle?

10 cm

6 cm

Workspace

Day 60:

What is the perimeter of this square?

8 m

Workspace

Day 61:

What is the perimeter of the rectangle?

Workspace

Day 62:

What is the area of the figure below?

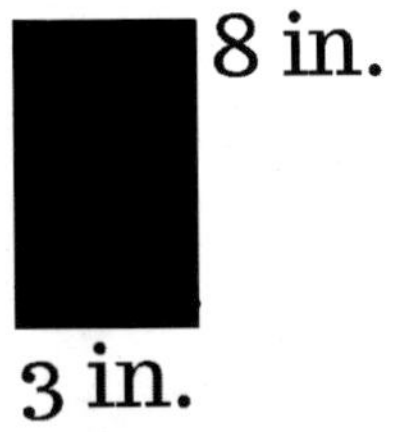

Workspace

Day 63:

Draw a right angle.

Workspace

Day 64:

Draw an obtuse triangle.

Workspace

Day 65:

The talent show costs $3.00 per person to attend. How much will it cost if 5 people go? 7 people? 9 people?

Week:	1	2	3	4	5	6
Cost:	$3	$6	$12	$15	$18	

Workspace

Day 66:

Reduce 8/12 to the lowest term.

Workspace

Day 67:

What two fractions are equivalent to 4/5?

Workspace

Day 68:

Divide this pizza into eight equal parts.

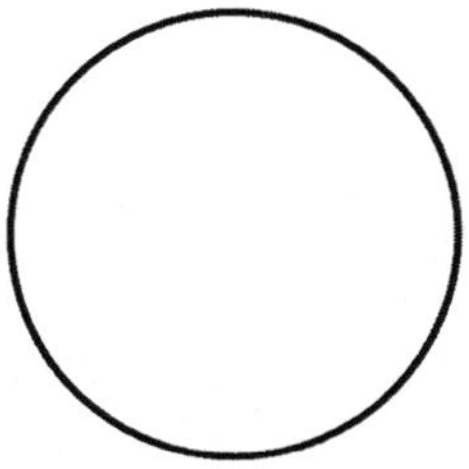

Shade in $5/8$ of the pizza

Workspace

Day 69:

The cat's birthday cake has $\frac{1}{4}$ cup of milk and $\frac{3}{4}$ cup of water. How much liquid is needed in the cake?

Workspace

Day 70:

James spent 3/6 of an hour studying for his social studies exam and 5/6 of an hour studying for his French test. How much longer did James spend studying for French than social studies?

He spent _______ of an hour more studying for French than studying social studies.

Workspace

Day 71:

Thea bought ¾ lbs. of green bananas and 2/4 lbs. of yellow (ripe) bananas did Thea buy in all?

Thea bought ________ lbs. of bananas in all.

Workspace

Day 72:

Helen baked 252 cookies. She wants to give all six of her grandchildren the same number of cookies. How many cookies would each grandchild get?

Helen will give each grandchild __________ cookies.

Workspace

Day 73:

Jason received $85.00 from his part-time job. He went to a football game, a movie, and the game room. He had $12.50 left. How much did he spend on entertainment?

Jason spent $__________ on entertainment.

Workspace

Day 74:

Last week, Ethan made $79.00 doing chores. He decided to wash cars at the Pit Stop. He now has $130.00. How much money did he make washing cars?

Jason made $__________ washing cars.

Workspace

Day 75:

Fredrick has 85 watermelons left at his vegetable stand. He sent his workers to get more watermelons to stock for the afternoon. He now has 105 watermelons at his stand. How many did the workers bring from the field?

Fredrick's workers brought ________ watermelons from the field.

Workspace

Day 76:

Draw a clock with the time 1:40.

- What time will it be in 2 hours and 30 minutes? ___________
- What time will it be in 3 hours and 15 minutes? ___________

Workspace

Day 77:

Are these lines below parallel? Explain

Workspace

Day 78:

Are the lines below parallel? Explain

Workspace

Day 79:

Are the lines below intersecting? Explain.

Workspace

Day 80:

This shape has how many vertices? Explain?

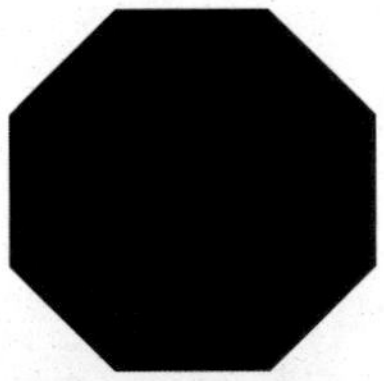

Workspace

Day 81:

How many sides does this shape have?

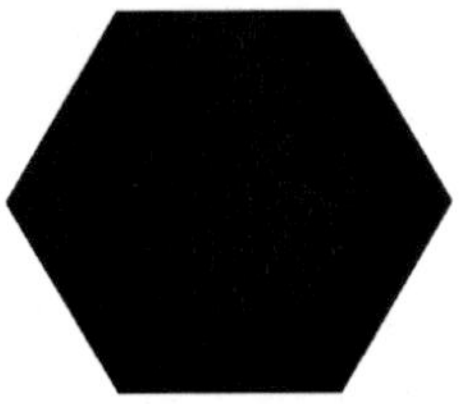

This shape has ____________ sides.

Workspace

Day 82:

What is the unknown number?

$7 + r = 11$

Workspace

Day 83:

Johnny has $24.00 to spend on six ink pens. Aft6er buying the pens, he has $12.00 left. How much did each ink pen cost?

East ink pen cost $__________

Workspace

Day 84:

There were 324 students going to the zoo. The students filled six buses. How many students were on each bus?

Workspace

Day 85:

After having lunch at the restaurant, Dee, Rosemary, and Robin decided to split the cost for lunch evenly. If each person paid $19.00, what was the total bill for lunch?

Workspace

Day 86:

Converting: How many minutes and hours are in 205 minutes?

Remember: 60 minutes = 1 hour

Workspace

Day 87:

Converting: How many minutes and hours are in 159 minutes?

Remember: 60 minutes = 1 hour

Workspace

Day 88:

Converting: How many minutes are in 5 hours?

Remember: 60 minutes = 1 hour

It is __________ minutes in 5 hours

Workspace

Day 89:

Solve the equation:

$16 = -10a + 2a$

Workspace

Day 90:

Solve the equation:

$3 = 6t - 12 - 3$

Workspace

Day 91:

Solve the equation:

$36 = -p - 5p$

Workspace

Day 92:

Solve the equation:

$q + 8 - 2 = 9$

Workspace

Day 93:

Make a bar graph with 7 classmates' favorite vegetables:

Favorite Vegetables:

squash	green beans	cabbage	carrots	cucumber

What is the most favorite food? ________________

What is the least favorite food? ________________

Workspace

Day 94:

Make a bar graph with 10 classmates' favorite subjects:

Favorite Subject:

Reading	Math	Science	Social Studies	Writing

Workspace

Day 95:

Make a bar graph using a cup of colorful marshmallows and create two questions about your graph.

Workspace

Day 96:

Solve the equation:

$4 \times 27 \div 3$

Workspace

Day 97:

Solve the equation:

(20 x 3) ÷ 3

Workspace

Day 98:

Solve the equation:

3.07 + 5.4 - 2.1

Workspace

Day 99:

Using decimals: The vet is 7.8 kilometer west of the gas station and 25.6 kilometer east of the beauty salon. How far apart are the gas station and the beauty salon?

Workspace

Day 100:

Using decimals: The mechanic shop is 71.6 miles west of the donut shop and 54.5 miles east of the mechanic shop. How far is the donut shop from the mechanic shop?

Workspace

Day 101:

Rachael is reading a mystery novel that has 581 pages. She wants to finish reading the novel in seven days. How many pages does she need to read each day?

Workspace

Day 102:

the Parent and Teacher Organization (PTO) bought 14 dozen boxes of lead pencils. Each box has 18 lead pencils. the PTO paid $6.00 per dozen. If the PTO sells eachl ead pencil for $0.50, how much profit will the PTO make?

Workspace

Day 103:

Jason and his friends play baseball on Saturday. The game begins at 2:30 p.m. It will take Jason and his friends 40 minutes to get to the field. Approximately what time do they need to leave to get to the baseball field, 30 minutes before the game begins?

Workspace

Day 104:

Michael's football practice began at 11:00 a.m. The team practiced offense for 65 minutes and 1 hour and 45 minutes on defense techniques. What time did Michael's practice end?

Workspace

Day 105:

Elizabeth went to the park at 2:35 p.m. She played in the sandbox for 30 minutes, the slides for 10 minutes, and the swings for 15 minutes. What time did Elizabeth leave the park?

Workspace

Day 106:

On Tuesday, Emmanuel had a Student Council meeting for 45 minutes followed by keyboarding music lessons for 1 hour and 10 minutes. If Emmanuel gets out of school at 3:15 p.m., what time was he finished with his music lesson?

Workspace

Day 107:

John bought 43 cases of carnation flowers. There were 35 carnations in each case. How many carnations did John buy?

Workspace

Day 108:

The school district has 65 buses. There are 39 students on each bus. Approximately how many students can the buses transport to school?

Workspace

Day 109:

At Elizabeth's birthday party, the girls ate 2 ½ pizzas and the boys ate 4 ¾ pizzas. How many pizzas did the boys and girls eat in all?

Workspace

Day 110:

Mr. Johnny bought 5 gallons of paint. He used only 1 ¾ gallon. How much paint does Mr. Johnny have left?

Workspace

Day 111:

My German Shepherd is 6 ½ years old, and my Cocker Spaniel is 2 ½ years younger than the German Shepherd.
How old is my Cocker Spaniel?

Workspace

Day 112:

Illustrate the decimal number 0.4.

Workspace

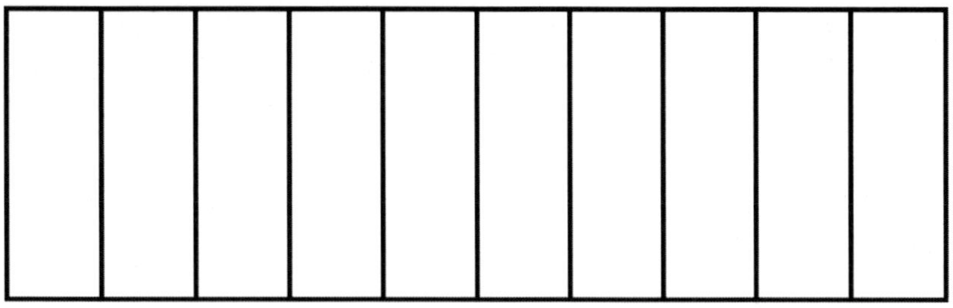

Day 113:

Illustrate the decimal number 0.74.

Workspace

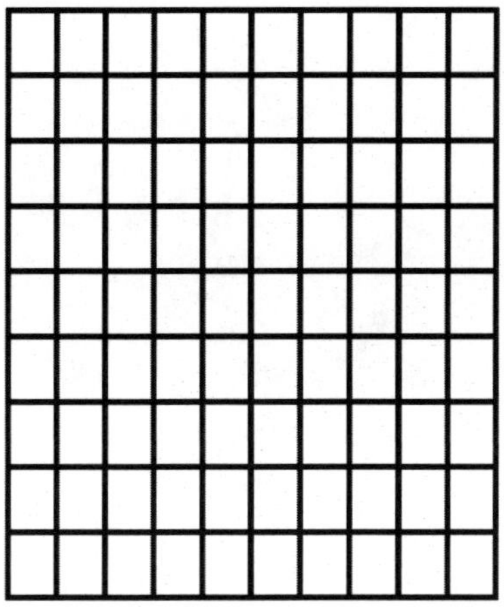

Day 114:

Illustrate the decimal number 0.04.

Workspace

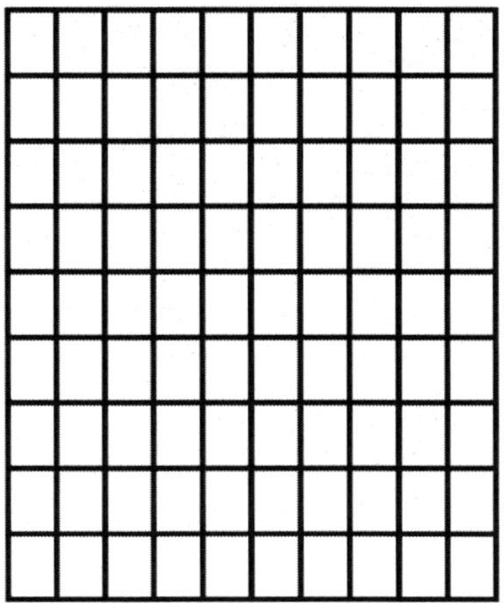

Day 115:

Janee bought 5.1 meters length of cloth to cover a scrapbook. She used 4.3 meters of the cloth. How much cloth did Janee have left?

Workspace

Day 116:

Jason had 0.9 grams of salt. He used only 0.5 grams of salt to make the chicken fritters. How much slat did Jason have left?

Workspace

Day 117:

Illustrate 3 blue marbles and 2 yellow marbles.
How likely is it that you will pick a blue marble out of a bag?
Explain.

Use words like certain, unlikely, impossible, and probably to solve problems.

Workspace

Day 118:

Illustrate: Five yellow rocks and 4 brown rocks.
How likely is it that you will pick a yellow rock?

Use words like certain, unlikely, impossible, and probably to solve problems.

Workspace

Day 119:

Place Value:

Looking at the number 3,485, which digit is in the tens place?

Workspace

Day 120:

Even Numbers:

Write the even number from 44 to ______

Workspace

44, ______, ______, ______, ______, ______, ______, ______, ______, ______,

______, ______, ______, ______, ______, ______, ______, ______, ______,

______

Day 121:

Odd Numbers:

Write the odd numbers from 1 to _____

Workspace

1, _____, _____, _____, _____, _____, _____, _____, _____, _____, _____, _____,

_____, _____, _____, _____, _____, _____, _____, _____, _____, _____, _____,

_____, _____, _____, _____, _____

Day 122:

Counting backwards by 2's

Workspace

50, _____, _____, _____, _____, _____, _____, _____, _____, _____,

_____, _____, _____, _____, _____, _____, _____, _____, _____,

_____, _____, _____, _____, _____

Day 123:

Counting by 4's from 44

Workspace

44, ____, ____, ____, ____, ____, ____, ____, ____, ____, ____,
____, ____, ____, ____, ____, ____, ____, ____, ____, ____, ____,
____, ____, ____, ____, ____, ____

Day 124:

Counting by 10's from 43.

Workspace

43, ____, ____, ____, ____, ____, ____, ____, ____, ____,
____, ____, ____, ____, ____, ____, ____, ____, ____,
____, ____, ____, ____, ____, ____, ____, ____, ____

Day 125:

Counting by 10's from 112.

Workspace

112, _____, _____, _____, _____, _____, _____, _____, _____, _____, _____,

_____, _____, _____, _____, _____, _____, _____, _____, _____, _____, _____,

_____, _____, _____, _____, _____, _____

Day 126:

Counting by 5's from 102.

Workspace

102, _____, _____, _____, _____, _____, _____, _____, _____, _____,

_____, _____, _____, _____, _____, _____, _____, _____, _____,

_____, _____, _____, _____, _____, _____, _____, _____, _____

Day 127:

Number words:
Write the number word for 587.

Workspace

Day 128:

Number words:
Write the number word for 1,802.

Workspace

Day 127:

Number words:

Write the number word for 587.

Workspace

Day 128:

Number words:

Write the number word for 1,802.

Workspace

Day 129:

Ethan bought a 5-Ib. bag of apples for $4.38, and he gave the cashier $5.00. How much change will he get?

Workspace

Day 130:

You went to the store to get a gallon of vanilla ice cream. The ice cream costs $5.89. Your mom gave you 3 one-dollar bills, 8 quarters, 5 dimes, and 6 pennies.

How much money did your mom give you? ________

Do you have enough money to get the ice cream?

Yes ______ No ________

How much money do you have left, or how much more do you need? __________

Workspace

Day 129:

Ethan bought a 5-Ib. bag of apples for $4.38, and he gave the cashier $5.00. How much change will he get?

Workspace

Day 130:

You went to the store to get a gallon of vanilla ice cream. The ice cream costs $5.89. Your mom gave you 3 one-dollar bills, 8 quarters, 5 dimes, and 6 pennies. how much money did your mom give you? ________

Do you have enough money to get the ice cream? Yes ______ No ________

How much money do you have left or how much more do you need? __________

Workspace

Day 131:

We are planting 48 sweet potato plants and would like for them to be in 7 rows. How many sweet potato plants need to be in each row?

Workspace

Day 132:

At your lemonade stand, you sold lemonade for $0.15 a cup. You made $18.00. How many cups of lemonade did you sell?

Workspace

Day 133:

The PTO sold 856 students T-shirts at school for $10.00 each. Everyone purchased a T-shirt for $10.00. How much money did the PTO make in selling the T-shirts? ____________

The PTO paid $6.25 for each T-shirt. How much profit did the PTO make?

Workspace

Day 134:

Mrs. Frederick made a batch of cookies. There were 24 cookies in each batch. She ate ¼ of the batch. Then, Mrs. Fredrick baked three more batches. How many cookies does Mrs. Fredrick have now?

Workspace

Day 133:

The PTO sold 856 students T-shirts at school for $10.00 each. Everyone purchased a T-shirt for $10.00. How much money did the PTO make in selling the T-shirts? ______________

The PTO paid $6.25 for each T-shirt. How much profit did the PTO make?

Workspace

Day 134:

Mrs. Frederick made a batch of cookies. There were 24 cookies in each batch. She ate ¼ of the batch. Then Mrs. Fredrick baked three more batches.

How many cookies does Mrs. Fredrick have now?

Workspace

Day 135:

Emmanuel's grades in science were 89, 93, 87, 77, 100, and 85. What is the average grade in science?

Workspace

Day 136:

Angie wants to give the server a 15% tip. Her lunch receipt is $18.89. How much will Angie give the server?

Workspace

Day 137:

A rectangular court measures 12 ft. x 5 ft.
What is the area of the court?

Workspace

Day 138:

A square room measures 8 ft. on the right side.
What is the perimeter?

Workspace

Day 139:

George is putting new carpet in his bedroom. His bedroom is 21 ft. x 17 ft. What is the area of his bedroom?

Workspace

Day 140:

Dr. Dargan has a rose garden that is 8 m by 12 m.
One bag of horse manure can cover 15 m.
How many bags of horse manure will Dr. Dargan use?

Workspace

Day 141:

Elapsed time: If it is 8:13 am., what time will it be in 7 hours and 15 minutes?

Workspace

Day 142:

Elapsed time: If it is 1:45 pm., what time will it be in 7 hours and 30 minutes?

Workspace

Day 143:

Mark started baking some chocolate chip cookies at 4:35 p.m. It took 35 minutes to prepare the ingredients and 2 hours and 10 minutes to bake the cookies. When did Mark finish baking?

Workspace

Day 144:

Ethan and Elizabeth were watching a video at 5:35 p.m. The video lasted for 1 hour and 45 minutes. Afterwards, Ethan and Elizabeth played basketball for 35 minutes. What time did Ethan and Elizabeth finish playing basketball?

Workspace

Day 145:

What shape has three equal sides?
Illustrate and name the shape.

Workspace

Day 146:

What shape has four equal sides and four equal corners?
Illustrate and name the shape.

Workspace

Day 147:

What shape has six sides and six angles?
Illustrate and name the shape.

Workspace

Day 148:

What shape has eight sides?
Illustrate and name the shape.

Workspace

Day 149:

Convert the number of miles it takes to equal 15,840 feet.

Workspace

Day 150:

How much time elapsed between the first and second time?

First	Second	Elapsed Time
1:45	2:30	____________
4:15	5:46	____________
9:35	9:45	____________

Workspace

Day 151:

Thea has 895 erasers in a box. She sold 186 erasers.
How many does she have left?

Workspace

Day 152:

Lilian has 834 dimes in her piggy bank. She spent 479 dimes.
How many dimes does she have now?

Workspace

Day 153:

Lula sold 819 cooking magazines on Monday, and sold an additional 749 magazines on Tuesday. How many more magazines did she sell on Monday than she sold on Tuesday?

Workspace

Day 154:

A toy factory made 589,877 trucks and 539,227 dolls. How many more trucks than dolls did the toy factory make?

Workspace

Day 155:

During the midnight shift, the workers made 5,277 small bobby pins and 3,456 large bobby pins. How many bobby pins did the workers make during the midnight shift?

Workspace

Day 156:

Last year, a candy factory made 6,899 chocolate Easter eggs. This year, the candy factory made 10,100 chocolate Easter eggs. How many more chocolate eggs did they make this year than it made last year?

Workspace

Day 157:

Simmons and Brown Enterprise sold 6,990 paperback books. The company also sold 3,011 eBooks. How many books did Simmons and Brown Enterprise sell in all?

Workspace

Day 158:

They cut down 4,660 trees on the north side of the estate. They cut down 5,896 more from the south side of the estate. How many more trees were cut down on the south side than on the North side?

Workspace

Day 159:

Wilhemenia wants to build a fence around her backyard.
Whilhemenia's backyard is 15 meters long and 15 meters wide.
It costs $7.00 per meter to install the fence.
How much will it cost to put fences around Wilhemenia's backyard?

Workspace

Day 160:

The CEO wants to install a new floor in the office. The office measures 9 meters wide and 12 meters long. The flooring costs $8.00 per square meter.
How much will the CEO pay for the new floor in the office?

Workspace

Day 161:

Patterns:

1, 3, 9, ____, 81, 243

1, 2, 4, ____, 16, 32

1, 3, 9, ____, 81

Workspace

Day 162:

The Ridgeville Community Resource Center used a grant to buy 13,456 books for preschool children. The center distributed 8,045 of the books. How many books does the Center have left?

Workspace

Day 163:

The staff at the Native American Center treated 879 clients this month. Last month, the staff treated 1,245 clients. How many more clients did the staff treat last month compared to this month?

Workspace

Day 164:

What is the median?

1) 3 5 5 7 8 9 9

2) 1 2 2 4 5 6 8

Workspace

Day 165:

What is the mean or average?

1) 1 6 7 3 8 9 3 3

2) 3 5 4 9 7 9 6 5

Workspace

Day 166:

What is the median?

1) 8 8 7 6 5 4 1

2) 9 6 4 7 8 3 2

Workspace

Day 167:

What is the mode?

1) 9 5 6 5 3 5 6 4

2) 0 0 7 9 2 1 0 3

Workspace

Day 168:

What is the mode?

1) 8 9 6 8 5 8

2) 7 7 7 9 9 3 3

Workspace

Day 169:

What is the range?

1) 5 9 4 8 9 4 3

2) 3 7 5 10 4 2 7 9

Workspace

Day 170:

Angie bought a bag of speckled beans. The clerk asked for $1.79 for the bag of speckled beans. What coins can she give the clerk to pay for the speckled beans?

Workspace

Day 171:

Wanda bought a container of eggs for $2.97. She gave the cashier $5.00. How much change will she get back?

Workspace

Day 172:

Jason bought a pack of fat-free cheese for $3.99. He has 2 dollars, 4 quarters, and 5 dimes. How much change will Jason get back or how much money will he need to buy the fatofree cheese?

Workspace

Day 173:

Write a word problem with your partner and design a strategy to get an answer.

Workspace

Day 174:

Write a word problem with your partner and design a strategy to get an answer.

Workspace

Day 175:

Write a word problem with your partner and design a strategy to get an answer.

Workspace

Day 176:

Write a word problem with your partner and design a strategy to get an answer.

Workspace

Day 177:

Write a word problem with your partner and design a strategy to get an answer.

Workspace

Day 178:

Write a word problem with your partner and design a strategy to get an answer.

Workspace

Day 179:

Write a word problem with your partner and design a strategy to get an answer.

Workspace

Day 180:

Write a word problem with your partner and design a strategy to get an answer.

Workspace

Answer Key

Day 1

It will take him 19 hours and 7 minutes to go 55 mph and 48 minutes to go 70 mph.

55 mph = 19 hours and 7 minutes

70 mph = 15 hours and 48 minutes

Day 2

Show and explain what you will do so that everyone has an equal slice of the pizza.

Day 3

Answer may vary.

Day 4

There are <u>6</u> boys in Dr. Now's class.

Day 5

Saturday School will end at 12:00 pm or noon.

Day 6

There were 8 cupcakes were left.

Day 7

50 + 50+ 50+ 50 = 200

or

25 X 8 = 200

I can pick 200 tomatoes in 8 hours

Day 8

289 X 4 = 1,156 legs

They have 1,156 legs altogether.

Day 9

Each potatoes chip has 11 calories.

Day 10

There are 61 blankets in each shipment.

Day 11

There are 24 rows of students in the marching band.

Day 12

1,858 bottles are filled every 30 minutes.

Day 13

588 ÷ 12 = 49

There are 49 rows of cement blocks.

Day 14

4 is level one X4

16 is level two X 4

64 is level three X 4

256 is level four X 4

256 X 4 = 1,024 is 5-level

There are 49 rows of 12 cement blocks that he can lay.

The rule is that the number of points scored on each subsequent level is multiplied by four.

Day 15

3 - 5 = 2

5 - 8 = 3

12 - 8 = 4

12 + 5 = 17

There are 17 marbles displayed in the fifth tray.

The rule is to add the next consecutive integer to the previous number.

Day 16

The rule being used in the pattern is adding 2 to the number of potatoes in the previous pile.

17 potatoes will be in the fifth pile.

Day 17

There are 56 barrettes in the sixth jar.

Day 18

5 centimeters is about equal to 2 inches.

Day 19

1 liter is about 1 quart.

Therefore, 5 liters is approximately 5 quarts.

Day 20

(29 - 5) + 29 = 53 inches in all.

Day 21

One century equals 100 years X.25 or ¼ = 25 years.

There are 25 years is in a fourth of a century.

Day 22

There are 6 minutes in 1/10 of an hour.

Day 23

There are 5 years in 1/2 of a decade.

Day 24

4/24 or 1/6 is a fraction of a day.

Day 25

10 - 8 = 2

2 + 16 = 18

10 +16 +2 = 28

She has 28 ribbons in total.

Day 26

He saved 54 + 35 = 89

He spent 10 + 15 + 14.75 +

25 = 64.75

$89.00 - 64.75 = $24.25

Ethan has $24.25 left.

Day 27

4+ 14 = 18

18 + 11 = 29

14 + 18 + 29 = 61

There are 61 makers in Michael's desk drawer.

Day 28

37 X 4 = 148

There are 148 fruit trees in his vineyard.

Day 29

816 ÷ 8 = 102

The Simmons Explorer will need to use 102 cabins.

Day 30

23 + 8 + 14 = 45

45 - 17 = 28

Elizabeth has 28 sparkling marbles left.

Day 31

Write an equation. Make a diagram.

123 X 3 = 369 feet

123 + 123 + 123 - 369

Look back and review.

Day 32

40 + 40 + 26 + 26 = 132 feet

The border of flowers needs to be 132 feet.

Day 33

18 + 18 + 12.5 + 12.5 = 61

Janee needs 61 inches of border.

Day 34

13 + 13 + 13 + 13 = 52

Your mother needs 52 yards of fence for her rose garden.

Day 35

69 X 85 = 5,865

The daycare ordered 5,865 Popsicles in all.

Day 36

654 ÷ 6 = 109

They need 109 sacks of potatoes.

Day 37

6,918.00 ÷3 = 2,459 tiles

The principle can buy 2,459 tiles.

Day 38

780 ÷ 15 = 52

The district needs 52 passengers.

Day 39

The squad can purchase approximately

$829 ÷ $ 85 = 975 buttons for their fans.

Day 40

The seniors will need approximately 554 ÷ 9 = 62 cars.

Day 41

He spent $287 + $878 = $1,165 in all on your mother's jewelry.

Day 42

Jason's science class starts at 10:35 a.m.

Day 43

$7\frac{1}{4} + \frac{3}{4} = 8$

$7 + 4/4 = 8$

$7 + 1 = 8$

You will need to drink 8 cups of water daily.

Day 44

$24 \div 4 = 6$

$6 \times 4 = 12$

$\frac{3}{4}$ of $24 = 12$

There are 12 boys in the class.

Day 45

1/8 of 48

$48 \div 8 = 6$

$6 \times 1 = 6$

There are six cupcakes with green icing.

Day 46

3/6 of 18

$18 \div 6 = 3$

$3 \times 3 = 9$

3/6 of 18 = 9

There are nine children playing kickball.

Day 47

¾ of 36

$36 \div 4 = 9$

$9 \times 3 = 27$

¾ of 36 = 27

27 students are eating in the cafeteria for lunch.

Day 48

1/5 of 50

$50 \div 5 = 10$; $1 \times 10 = 10$; 1/5 of 50 = 10

10 people are waiting with more than two pieces of luggage.

Day 49

895 ÷ 5 = 179

X 179 = 179

1/5 of 895 = 179

179 students are wearing glasses.

Day 50

750 ÷ 5 =

150

150 X 2 = 300

2/5 of the 750 houses are brick homes, which equals 300.

Day 51

$12.00 X 6 = $48

48.00 X 4 = $192

or

48 + 48 + 48+ 48 = $192

They will pat $192 to rent the bikes for 6 hours.

Day 52

$25.00 + $35.00 = $60.00

Jason earned $60.00 in all

Day 53

30 + 25 = 55

55 minutes X 7 days = 385

Janee walked and did aerobics for 385 minutes or 6 hours and 25 minutes.

Day 54

$12 X 9 = $108

$15 X 6 = $90

$108 + $90 =$198

It would cost $198 for six adults and nine children to attend the amusement park.

Day 55

Michael needs to leave his house by 7:23 a.m. to get to work on time.

Day 56

\$59.00 + \$29.99 = \$88.99

\$88.99 + \$1.01 = \$90.00

Mrs. Simmons gave the clerk \$90.00.

Day 57

$4/12 \div (4/4) = 1/3$

This fraction in its lowest term is 1/3.

Day 58

$6/8 \div 2/2 = 3/4$

This fraction in its lowest term is ¾

Day 59

$A = l \times w$

The area of a rectangle is 10 cm x 6 cm = 60 cm.

Day 60

$P = 4s$

$P = 4 \times 8$

The perimeter of a square is 32 m.

Day 61

P = 2l + 2w

P = (2x6) + (2x7)

P = 12 + 14

P = 26 cm

The perimeter of the rectangle is 26 cm.

Day 62

The area of the figure is

A = l x w

A = 8 x 3

A = 24 in

Day 63

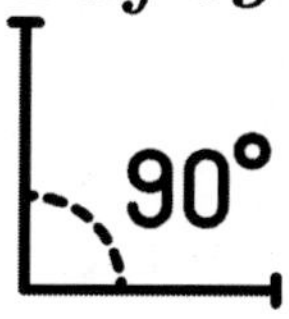

Day 64

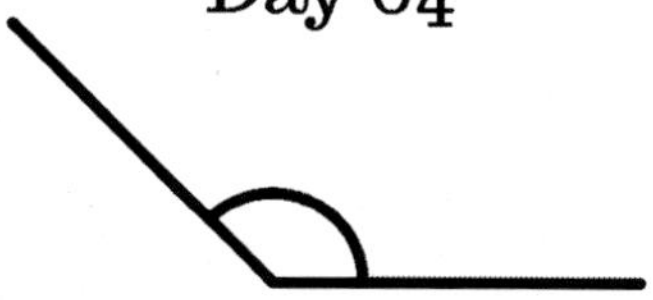

Day 65

Week 1 2 3 4 5 6

Cost $3 $6 $9 $12 $15 $18

$3 X 5 = $15

$3 X 7 = $21

$3 X 9 = $27

Day 66

8/12 ÷ 4/4 =2/3

The fraction in its lowest term is 2/3.

Day 67

4/5 = 8/10

= 16/20

The two fractions equivalent to 4/5 are 16/20 and 8/10.

Day 68

Shade in 5/8 of the pizza.

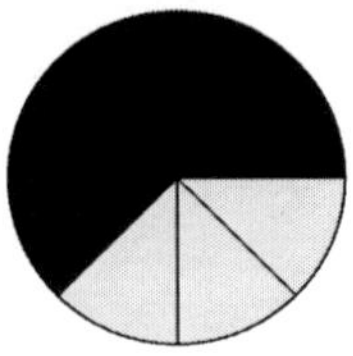

Day 69

¼ + ¾ = 4/4 = 1

1 cup of liquid is needed in the cake.

Day 70

5/6 - 3/6 = 2/6

2/6 = 1/3

He spent 1/3 of an hour more studying French than studying social studies.

Day 71

¾ + 2/4 =

5/4 = 1 ¼

Thea bought 1 ¼ lbs. of bananas in all.

Day 72

Helen will give each grandchild 42 cookies.

252 ÷ 6 = 42

Day 73

$ 85.00

-$ 12.50

$ 72.50

Jason spent $72.50 on entertainment.

Day 74

$135.00

- $79.00

―――――

$ 56.00

Jason made $56.00 washing cars.

Day 75

105

- 85

20

Frederick's workers brought 20 watermelons from the field.

Day 76

a. 4:10

b. 4:55

c. What time will it be in 2 hours and 30 minutes? ____

d. What time will it be in 3 hours and 15 minutes? ____

Day 77

No. Parallel lines stay the same distance apart. They never intersect.

Day 78

Yes. Parallel lines stay the same distance apart.

Day 79

Yes. The lines meet at one point.

Day 80

Vertices are the same as corners, 8 vertices.

Day 81

Sides are the straight lines that make up a shape.

This shape has 6 sides.

Day 82

7 + r = 11

7 + r -7 = 11 -7

r = 4

The unknown number is 4.

Day 83

$24 - 12 = 12

12 ÷ 6 = $2

Each ink pen costs $ $2.

Day 84

324 ÷ 6 =54

54

X 6

324

There were 324 students on each bus.

Day 85

$19

X 3

$57

$57 is the total bill for dinner.

Day 86

	3	Check
60	205	60 x 3= 180
	- 180	
	25	160 +25=205

There are 3 hours and 25 minutes in 205 minutes.

Day 87

60 minutes = 1 hour

Check:

$60 \times 2 = 120$

$$\begin{array}{r} 159 \\ -120 \\ \hline 39 \end{array}$$

$120 + 39 = 159$

2 hours and 39 minutes

Day 88

60 X 5

300 minutes

It is 300 minutes in 5 hours.

Day 89

$16 = -10a + 2a$

$16 = -8a$

$16/-8 = 8a/8$

$-2 = a$

Solve the equation: $16 = -10a + 2a$

Day 90

$3 = 6t - 12 - 3$

$3 = 6t - 15 + 15$

$3 + 15 = 6t$

$18 = 6t$

$18/6 = 6t / 6$

$3 = 6t - 12 - 3 = 3 = t$

Day 91

$36 = -p - 5p$

$36 = -6p$

$36 / 6 = -6p/-6$

$-6 = p$

Day 92

$q + 8 - 2 = 9$

Work Space:

$q + 8 - 2 = 9$

$q + 6 = 9$

$q + 6 - 6 = 9 - 6$

$q = 3$

Day 93

Make a bar graph with seven classmates' favorite vegetables.

What is the favorite food? ____

What is the least favorite food? ______

Answers will vary

Day 94

Create a graph of 10 classmates' favorite subjects:

Answers may vary

Day 95

Make a bar graph using a cup of colorful marshmallows and create two questions about your graph.

Day 96

$4 \times 27 \div 3 =$

$4 \times 9 =$

36

Day 97

Work Space:

$(20 \times 3) \div 3$

$(60) \div 3$

$60 \div 3 = 20$

$(20 \times 3) \div 3$

Day 98

$3.07 + 5.4 + 2.1 = 10.57$

Day 99

$25.6 - 7.8 = 17.8$

The gas station is 17.8 miles away from the beauty salon.

Day 100

71.6
-54.5
______
17.1

The donut shop is 17.1 miles from the mechanic shop.

Day 101

$$\begin{array}{r} 83 \\ 7\overline{)581} \\ -56 \\ \hline 21 \\ -21 \end{array}$$

Check:

83
x7
______
581

Rachael needs to read 83 pages each day.

Day 102

Pencils: 252

Profit: $42.00

Day 103

1:25 pm

Day 104

Practice ends at 1:50 p.m.

Day 105

Elizabeth left the park at 3:30 p.m.

Day 106

He was finished at 5:10 p.m.

Day 107

43 X 35 = 1,505

John bought 1,hzc505 carnations.

Day 108

65 X 39 = 2,535

The buses can transport 2,535 students.

Day 109

1 ½ = 2/4

4 ¾ = ¾

The boys and girls ate 7 ¼ pizzas.

6 5/4

7 ¼

Day 110

Mr. Johnny 3 ¼ gallons of paint left.

5	=	4 4/4
-1 3/4	=	1 3/4
———	=	3 1/4

Day 111

My Cocker spaniel is 4 years old.

6 ½

-2 ½

———

4

Day 112

Illustrate the decimal number 0.4.

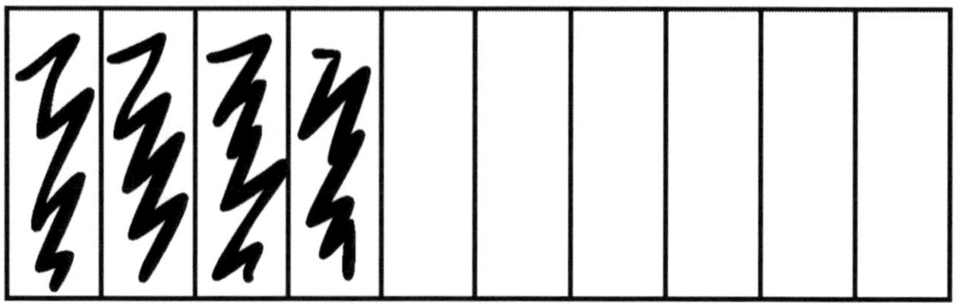

Day 113

Illustrate the decimal number 0.74.

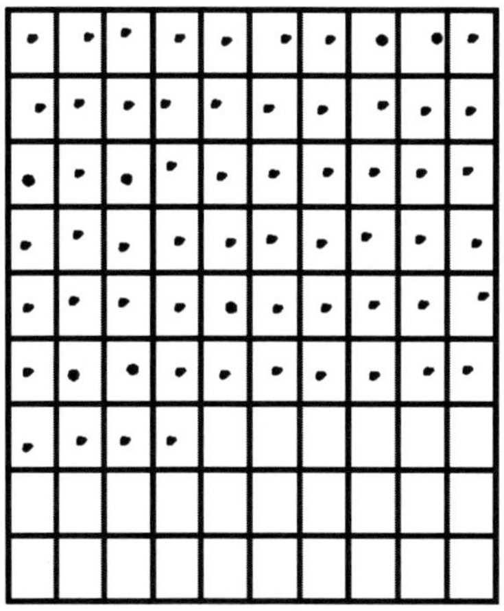

Day 114

Illustrate the decimal number 0.04.

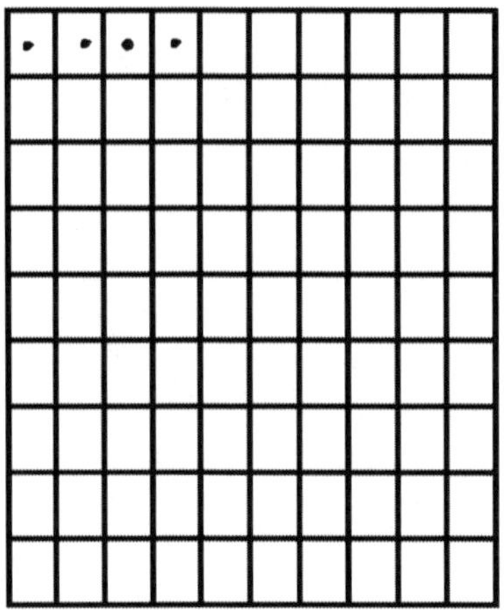

Day 115

5.1

-4.3

0.8

Janee had 0.8 of the cloth left.

Day 116

0.9

-0.5

0.4

Jason had 0.4 grams of salt left.

Day 117

It is probably likely

Day 118
It is Certain.

Day 119
8 is in the tens place.

Day 120
Even Numbers:
44, 46, 48, 50, 52, 54, 56, 58, 60, 62, 62, 66, 68, 70, 72, 74, 76,

Day 121
Odd Numbers: 1, 3, 5, 7, 9, 11, 13, 15, 17, 19, 21, 23, 25, 27, 29, 31, 33, 35, 37, 39, 41, 43, 45, 47, 49, 51, 53, 55, 57

Day 122
Counting backward by 2's: 50, 48, 46, 44, 42, 40, 38, 36, 34, 32, 30, 28, 26, 24, 22, 20, 18, 16, 14, 12, 10, 8, 6, 4, 2, 0

Day 123
Counting by 4's: 44, 48, 52, 56, 60, 64, 68, 72, 76, 80, 84, 88, 92, 96, 100, 104, 108, 112, 116, 120, 124, 128, 132, 136, 140, 144, 148, 152

Day 124
Counting by 10s starting: 43, 53, 63, 73, 83, 93, 103, 113, 123, 133, 143, 153, 163, 173, 183, 193, 203, 213, 223, 233, 243, 253, 263, 273, 283, 293, 303.

Day 125
Counting by 10's: 112, 122, 132, 142, 152, 162, 172, 182, 192, 202, 212, 222, 232, 242, 252, 262, 272, 282, 292, 302, 312, 322, 332, 342, 352, 362, 372, 382, 392.

Day 126
Counting by 5's: 102, 107, 112, 117, 122, 127, 132, 137, 142, 147, 152, 157, 162, 167, 172, 177, 182, 187, 192, 197, 202, 207, 212, 217, 222, 227, 232, 237.

Day 127

587 written five hundred eighty-seven

Day 128

1,802 written one thousand eight hundred two

Day 129

$$\begin{array}{r} \$5.00 \\ -4.38 \\ \hline \$0.62 \end{array}$$

Ethan will get $.62.

Day 130

Mom gives you $5.56.

No, I do not have enough money to get the ice cream.

How much money do you have left, $.33?

$$\begin{array}{r} \$3.00 \\ 2.00 \\ 0.50 \\ +0.6 \\ \hline \$5.56 \end{array} \qquad \begin{array}{r} \$5.89 \\ -5.56 \\ \hline \$0.33 \end{array}$$

Day 131

6 rows will have seven potato plants, and 1 row will have six plants (answer may vary)

Day 132

\$18.00 / \$0.15 = 120 cups

Day 133

856 X 10 = \$8,560 made.

6.25 X 856 = \$5,350 paid.

\$8,560

- 5,350

\$3,210 profit

PTO made \$8,560 from T-shirts. PTo made a \$3,210 profit.

Day 134

24 x 4 = 96

24 ÷ ¼ = 6

96 - 6 = 90

Mrs. Frederick has 90 cookies now.

Day 135

The average grade in science is 88.5.

Day 136

18.89 x .15 = \$ 2.83

Day 137

A = l x w

12 x 5 = 60 sq. ft

60 sq.ft is the area of the court.

Day 138

P = 8 + 8 + 8 + 8

A square has four equal sides. The perimeter is 32 sq. ft.

Day 139

A = l X w

21 X 17 = 147

357 sq. ft. is the area of his bedroom.

Day 140

12 x 8 = 96 m

96 ÷ 15 = 6 bags

Mr. Dargan will use 4 to 6 bags of horse manure.

Day 141

It will be 3:28 p.m.

Day 142

It will be 9:15 p.m.

Day 143

Mark finished baking at 7:15 p.m.

Day 144

Elizabeth and Ethan finished playing basketball at 7:55 p.m.

Day 145
Equilateral triangle

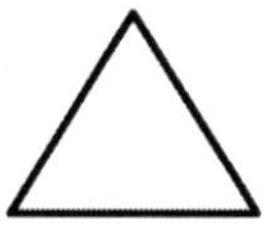

Day 146
Square

Day 147
Hexagon

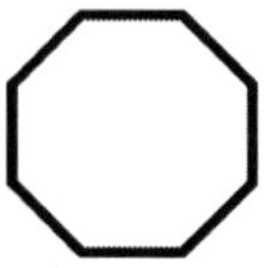

Day 148
Octagon

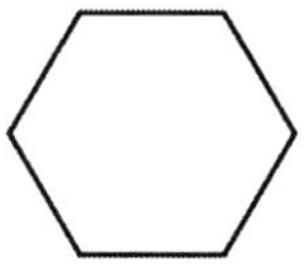

Day 149
1 mile = 5,280 feet
5,280 x 3 = 15,840 feet
3 miles is equal to 15,840 feet.

Day 150
45 minutes
1 hour 31 minutes
10 minutes

Day 151
895
-186
709
Thea has 709 erasers left.

Day 152
834
-479
———
355
Lillian has 355 dimes left.

Day 153

819-749= 70

Lula sold 70 more magazines on Monday than she did on Tuesday.

Day 154

589,877 - 539,227 = 50,650

The factory made 50,600 more trucks than dolls.

Day 155

5,277 + 3,456 = 8,733

The workers made 8,733 bobby pins.

Day 156

10,100 - 6,899 = 3,201

The Factory made 3,201 more eggs than they made last year.

Day 157

6,990

+3,011

10,001

Simmons and Brown Enterprise sold 10,001 in all.

Day 158

5,896

-4,660

1,236

There were 1,236 more trees cut down on the south side than on the north side.

Day 159

15 X 15 = 225

225 X $ 7.00 = $1,575

It costs $1,575 to put a fence around Wilhemenia's backyard

Day 160

12 X 9 = 108

108 X $8.00 = $864

The CEO will pay $864 for the new floor in the office.

Day 161

Patterns:

1, 3, 9, 27, 81, 243

1, 2, 4, 8, 16, 32

1, 3, 9, 27, 81

Day 162

13,456 - 8,045 =5,411

Day 163

1245 - 879 = 366

The staff have 366 compared to last month.

Day 164

1) 3 5 5 7 8 9 9

7 is the median

2) 1 2 2 4 5 6 8

4 is the median

When numbers are arranged from the least to the greatest, the median is the number in the middle.

Day 165

The mean is

1) 1 + 6 + 7 + 3 + 8 + 9 + 3 + 3 = 40 ÷ 8 = 5

2) 3 + 5 + 4 + 9 + 7 + 9 + 6 + 5 = 48 ÷ 8 = 6

The mean is the average of the numbers.

Day 166

What is the median?

1) 1 4 5 6 7 8 8

6 is the median

2) 2 3 4 6 7 8 9

6 is the median

The median is the number in the middle; therefore, the numbers must be arranged from the least to the greatest. Then find the number in the middle.

Day 167

1) 5

2) 0

Mode is the number that appears the most.

Day 168

1) 8

2) 7

Mode is the number that appears the most.

Day 169

1) 9 - 3 = 6

2) 10 - 2 = 8

The range is the difference between the largest and the smallest number.

Day 170

The coins will vary.

Day 171

$5.00

-2.97

$2.03

Wanda will get $2.03 back.

Day 172

$2.00 + $1.00 + $.50 = $3.50

$3.99

-3.40

$0.44

Jason will get back $0.44.

Day 173
Answers may vary.

Day 174
Answers may vary.

Day 175
Answers may vary.

Day 176
Answers may vary.

Day 177
Answers may vary.

Day 178
Answers may vary.

Day 179
Answers may vary.

Day 180
Answers may vary.

About the Author

Dr. Rosemary Simmons-Brown (a.k.a. Dr. Now) is the Director of the Reaching Your Dream Learning Center (RYDLC). She is a well-regarded educational leader from Dorchester County, South Carolina. Dr. Now has dedicated over 37 years to teaching and academic leadership in the public school system. Her commitment to students is evident in her role as parent-teacher facilitator and her authorship of mathematics skill enhancement books, including the Primary Word Problems series.

Dr. Now is the co-founder of the RYDLC. This learning center supports children from low socioeconomic backgrounds by providing a nurturing environment that promotes curiosity and learning. Dr. Simmons-Brown's educational journey began with a high school diploma in 1975 and culminated in a Doctorate in Educational Leadership from Nova Southeastern University in 2011. She resides in Ridgeville, SC, with her devoted husband, Rev. Johnny Henry Brown.

Made in United States
North Haven, CT
05 November 2026

21772053R00072